Get to Know Big Cats

LYNXES

By Bray Jacobson

Gareth Stevens
PUBLISHING

Please visit our website, www.garethstevens.com. For a free color catalog of all our high-quality books, call toll free 1-800-542-2595 or fax 1-877-542-2596.

Library of Congress Cataloging-in-Publication Data
Names: Jacobson, Bray, author.
Title: Lynxes / Bray Jacobson.
Description: Buffalo, New York : Gareth Stevens Publishing, [2024] | Series: Get to know big cats | Includes index. | Audience: Grades K-1
Identifiers: LCCN 2022045108 (print) | LCCN 2022045109 (ebook) | ISBN 9781538286111 (library binding) | ISBN 9781538286104 (paperback) | ISBN 9781538286128 (ebook)
Subjects: LCSH: Lynx (Genus)–Juvenile literature.
Classification: LCC QL737.C23 J3374 2024 (print) | LCC QL737.C23 (ebook) | DDC 599.75/3–dc23/eng/20220923
LC record available at https://lccn.loc.gov/2022045108
LC ebook record available at https://lccn.loc.gov/2022045109

First Edition

Published in 2024 by
Gareth Stevens Publishing
2544 Clinton Street
Buffalo, NY 14224

Editor: Kristen Nelson
Designer: Leslie Taylor

Photo credits: Cover, p. 1 Vaclav Matous/Shutterstock.com; p. 5 Tomas Hulik ARTpoint/Shutterstock.com; pp. 7, 24 (fur) Lucie Bartikova/Shutterstock.com; pp. 9, 24 (tail) Kjetil Kolbjornsrud/Shutterstock.com; p. 11 Geoffrey Kuchera/Shutterstock.com; p. 13 Volodymyr Burdiak/Shutterstock.com; p. 15 Jan Stria/Shutterstock.com; p. 17 Pictureguy/Shutterstock.com; p. 19 slowmotiongli/Shutterstock.com; p. 21 Bildagentur Zoonar GmbH/Shutterstock.com; pp. 23, 24 (kitten) Holly Kuchera/Shutterstock.com.

Printed in the United States of America

CPSIA compliance information: Batch #CSGS24: For further information contact Gareth Stevens at 1-800-542-2595.

Contents

Lynxes are big cats!

Their fur is thick.
It is light colored.
It has black or brown
spots too.

They have a short tail.

There are four kinds.
One kind is called
a bobcat.

Most live in places with snow.

These cats have long legs.
They have big paws.
They walk through snow!

They live alone.
They hunt at night.

Some hunt rabbits. Some hunt deer and moose.

Babies are kittens.
They are born in spring.

Their eyes are closed at birth.

Words to Know

fur

kitten

tail

Index

SNAKES Play Tricks!

BY JENNIFER LOMBARDO

ANIMAL TRICKSTERS

Gareth Stevens PUBLISHING

Please visit our website, www.garethstevens.com. For a free color catalog of all our high-quality books, call toll free 1-800-542-2595 or fax 1-877-542-2596.

Cataloging-in-Publication Data

Names: Lombardo, Jennifer.
Title: Snakes play tricks! / Jennifer Lombardo.
Description: New York : Gareth Stevens Publishing, 2025. | Series: Animal tricksters | Includes glossary and index.
Identifiers: ISBN 9781538293577 (pbk.) | ISBN 9781538293584 (library bound) | ISBN 9781538293591 (ebook)
Subjects: LCSH: Snakes–Juvenile literature. | Snakes–Behavior–Juvenile literature.
Classification: LCC QL666.O6 L66 2025 | DDC 597.96–dc23

Published in 2025 by
Gareth Stevens Publishing
2544 Clinton Street
Buffalo, NY 14224

Designer: Andrea Davison-Bartolotta
Editor: Jennifer Lombardo

Photo credits: Cover, p. 1 reptiles4all/Shutterstock.com; series art (background texture) Kriangkrai Namtongbai/Shutterstock.com; series art (scale texture) vuvu/Shutterstock.com; p. 5 Julian Popov/Shutterstock.com; p. 7 (globe) tele52/Shutterstock.com; p. 7 (snake) EcoPrint/Shutterstock.com; pp. 9 (bottom), 21 (bottom) Kurit afshen/Shutterstock.com; p. 9 (top) YuLu/Shutterstock.com; p. 11 Wirestock Creators/Shutterstock.com; p. 12 IvanRiv/Shutterstock.com; p. 13 (main) Colinmthompson/Shutterstock.com; p. 13 (inset) Gorlsko/Shutterstock.com; p. 15 Ken Griffiths/Shutterstock.com; p. 16 SAHAI IMAGE/Shutterstock.com; p. 17 Fly Dragon Fly/Shutterstock.com; p. 18 Mike Wilhelm/Shutterstock.com; p. 19 (inset) Mark_Kostich/Shutterstock.com; p. 19 (main) Amanda Guercio/Shutterstock.com; p. 21 (top) Steve Oehlenschlager/Shutterstock.com.

Printed in the United States of America

CPSIA compliance information: Batch #CSGS25: For further information contact Gareth Stevens at 1-800-542-2595.